SACRAMENTO PUBLIC LIBRARY
828 "I" Street
Sacramento, CA 95814
03/22

Blastoff! Readers are carefully developed by literacy experts to build reading stamina and move students toward fluency by combining standards-based content with developmentally appropriate text.

LEVELS

Level 1 provides the most support through repetition of high-frequency words, light text, predictable sentence patterns, and strong visual support.

Level 2 offers early readers a bit more challenge through varied sentences, increased text load, and text-supportive special features.

Level 3 advances early-fluent readers toward fluency through increased text load, less reliance on photos, advancing concepts, longer sentences, and more complex special features.

★ **Blastoff! Universe**

Reading Level

Grade K

Grades 1–3

Grade 4

This edition first published in 2022 by Bellwether Media, Inc.

No part of this publication may be reproduced in whole or in part without written permission of the publisher. For information regarding permission, write to Bellwether Media, Inc., Attention: Permissions Department, 6012 Blue Circle Drive, Minnetonka, MN 55343.

Library of Congress Cataloging-in-Publication Data

Names: Pettiford, Rebecca, author.
Title: Submarines / by Rebecca Pettiford.
Description: Minneapolis, MN : Bellwether Media, 2022. | Series: Blastoff readers! How it works | Includes bibliographical references and index. | Audience: Ages 5-8 | Audience: Grades 2-3 | Summary: "Simple text and full-color photography introduce beginning readers to submarines. Developed by literacy experts for students in kindergarten through third grade"-- Provided by publisher.
Identifiers: LCCN 2021049236 (print) | LCCN 2021049237 (ebook) | ISBN 9781644876022 (library binding) | ISBN 9781648346774 (paperback) | ISBN 9781648346132 (ebook)
Subjects: LCSH: Submarines--Juvenile literature.
Classification: LCC VM365 .P47 2021 (print) | LCC VM365 (ebook) | DDC 623.825/7--dc23/eng/20211012
LC record available at https://lccn.loc.gov/2021049236
LC ebook record available at https://lccn.loc.gov/2021049237

Text copyright © 2022 by Bellwether Media, Inc. BLASTOFF! READERS and associated logos are trademarks and/or registered trademarks of Bellwether Media, Inc.

Editor: Betsy Rathburn   Series Design: Jeffrey Kollock   Book Designer: Gabriel Hilger

Printed in the United States of America, North Mankato, MN.

Table of Contents

What Are Submarines? 4
How Do Submarines Work? 6
The Future of Submarines 20
Glossary 22
To Learn More 23
Index 24

What Are Submarines?

Submarines are machines that go underwater. They are often called subs. Subs can dive hundreds of feet!

Subs do many jobs. The military uses them for underwater work. Scientists use them to study the ocean.

research submarine

military submarine

How Do Submarines Work?

outer hull

Subs have two **hulls**. The long, narrow outer hull is waterproof. It keeps water out.

The inner hull holds the **crew**. It is very strong. It stands up against **pressure** from water. This keeps the crew safe!

Parts of a Submarine

- periscope
- sail
- outer hull
- propeller
- diving plane
- rudder

Subs use engines to run. Most engines can only run when a sub is at the surface. The running engines charge the sub's **battery**.

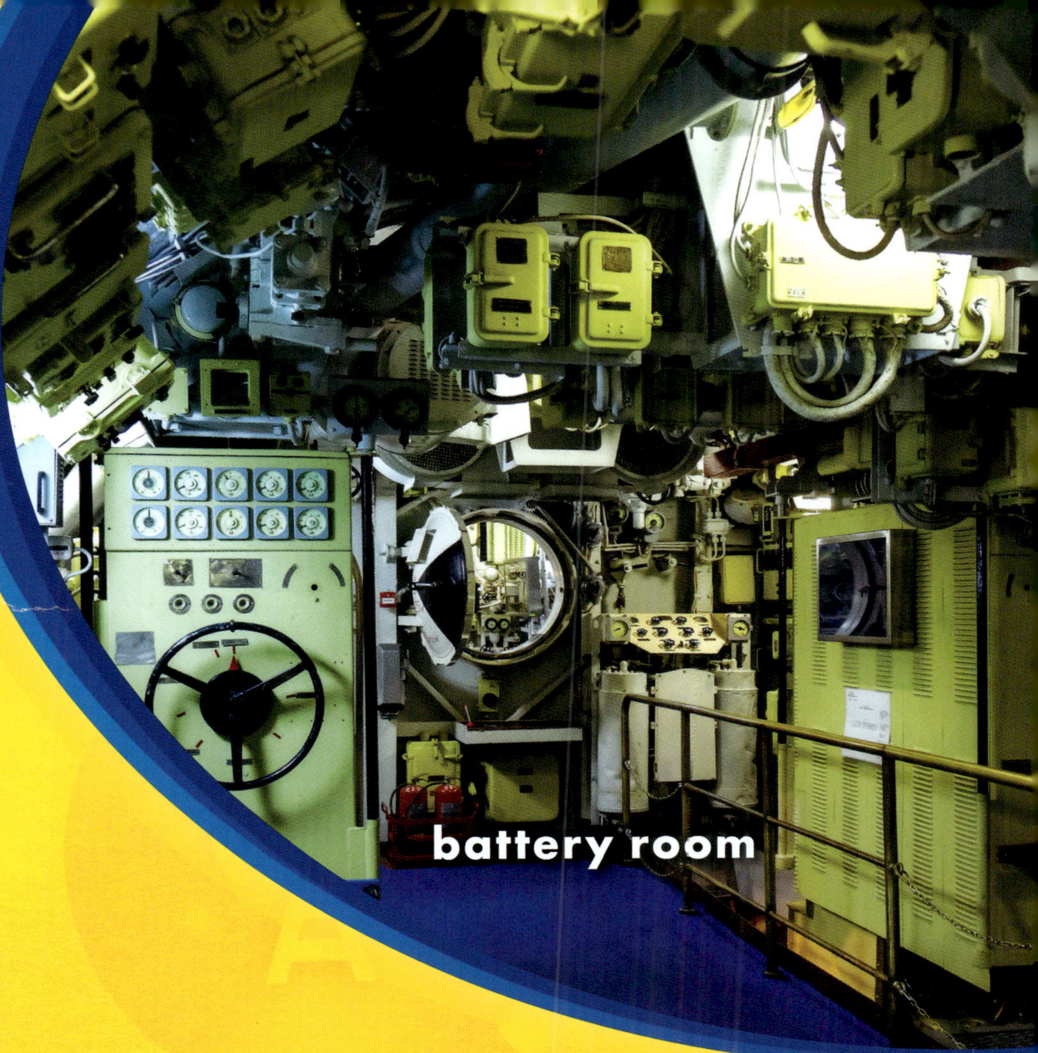

battery room

The battery runs the sub's other parts!

When a sub goes underwater, its engines shut down. The battery provides power.

The battery runs the **propeller**. Its spinning blades push the sub through water. **Rudders** help steer. **Diving planes** move the sub up and down.

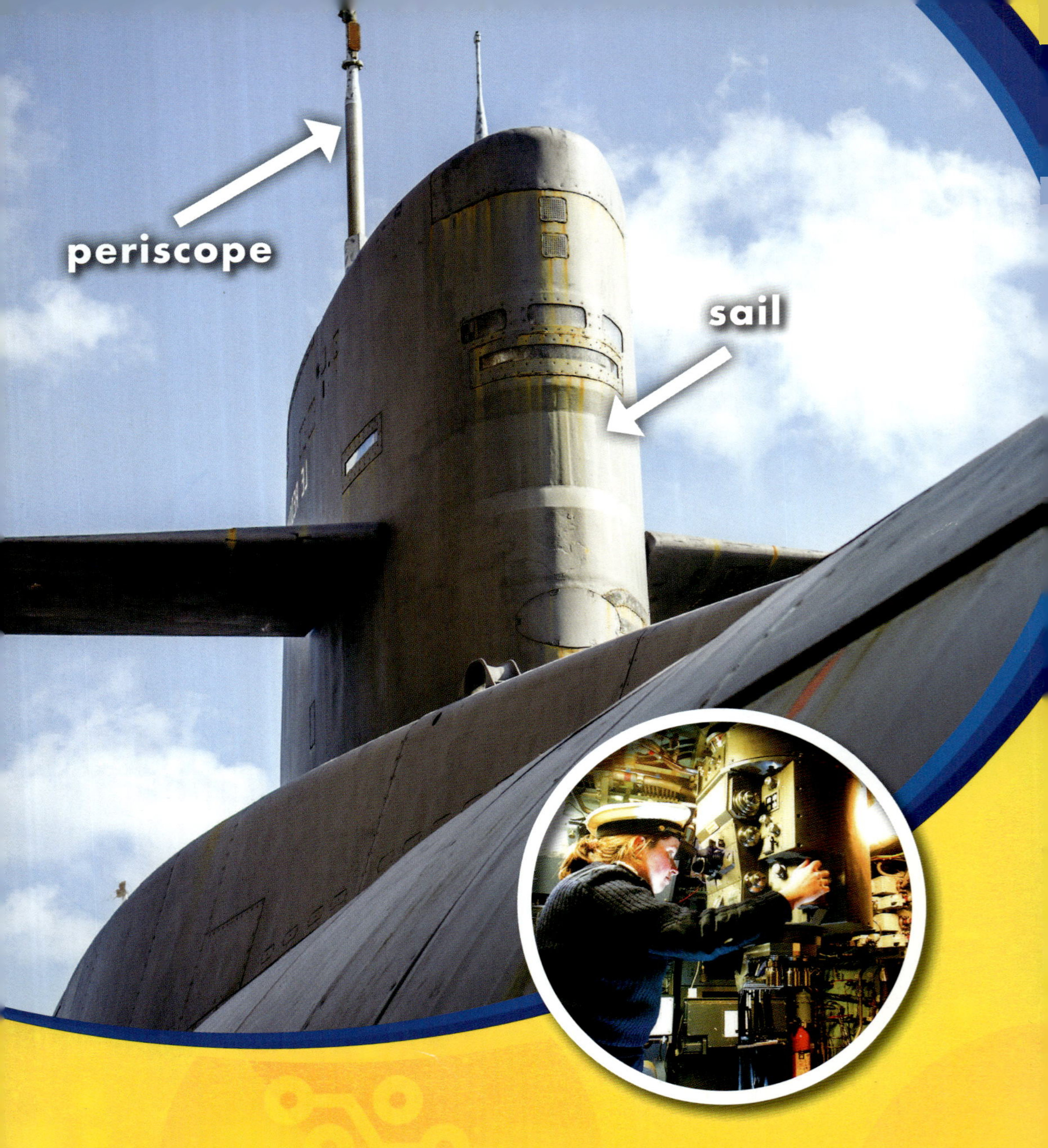

periscope

sail

Subs have a tall **sail**. It holds the **periscope**. The periscope rises above the water.

It helps the crew see what is at the surface. It also **communicates** with other subs!

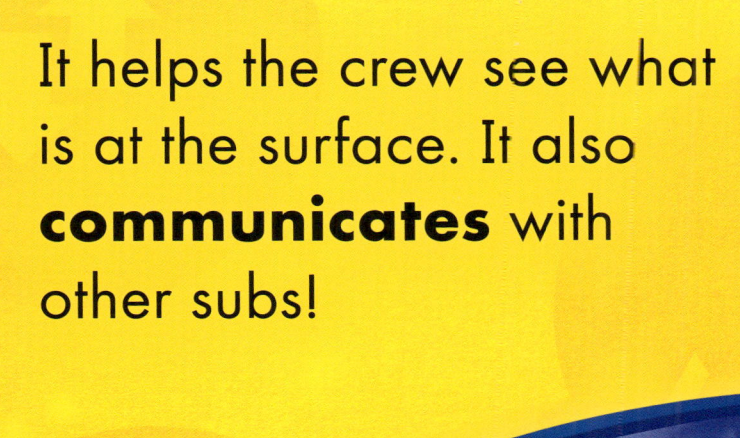

Subs are very heavy. But they are filled with air. They carry huge **ballast tanks**. Ballast tanks are full of air, too.

The air makes subs less **dense** than water. This makes them float!

ballast tank controls

To dive, subs fill their ballast tanks with water. This makes them dense enough to sink.

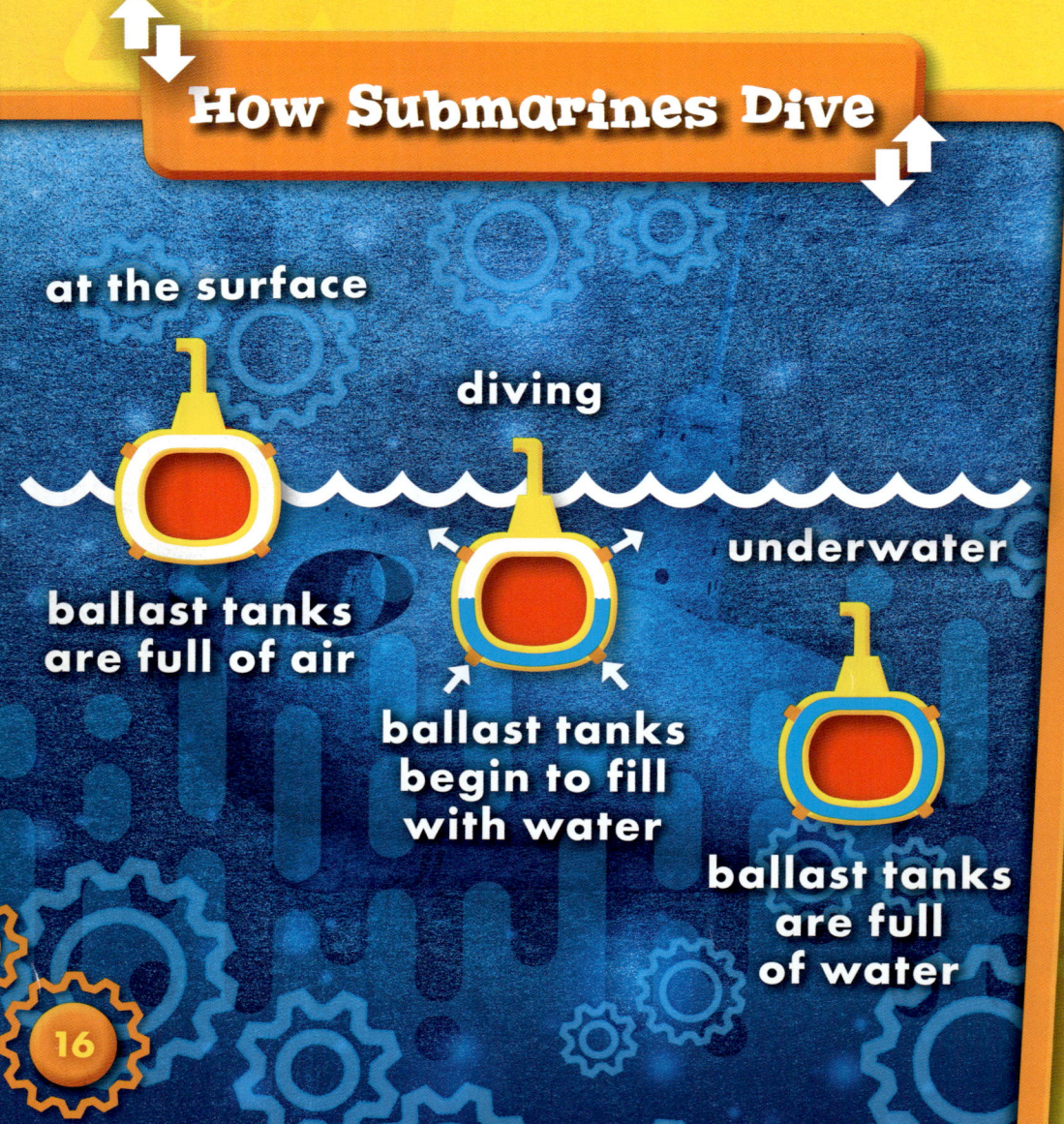

How Submarines Dive

at the surface — ballast tanks are full of air

diving — ballast tanks begin to fill with water

underwater — ballast tanks are full of water

Subs carry air with them. To come back up, they fill their ballast tanks with air!

Crews often cannot see what is ahead. Instead, they use **sonar**.

crew

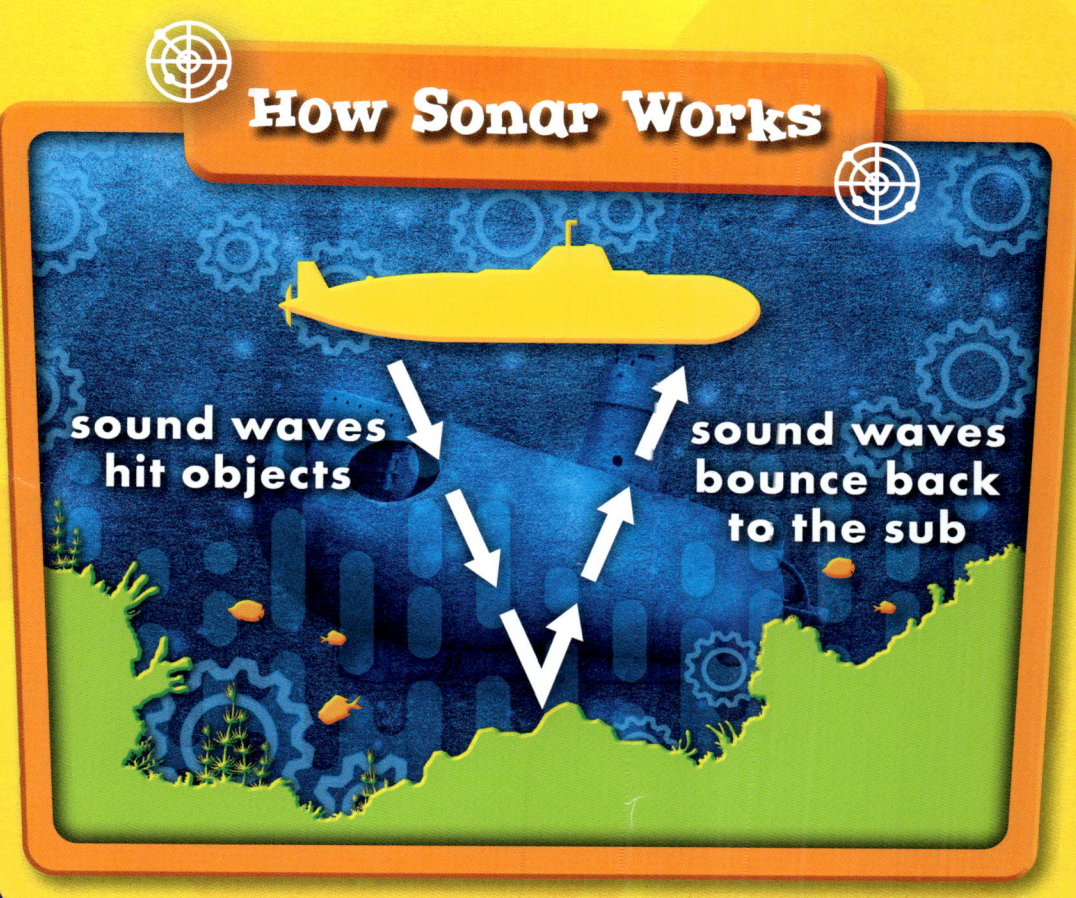

How Sonar Works

sound waves hit objects

sound waves bounce back to the sub

Sonar sends sound waves through the water. The waves bounce off objects they hit. This tells the crew how far away things are from the sub!

The Future of Submarines

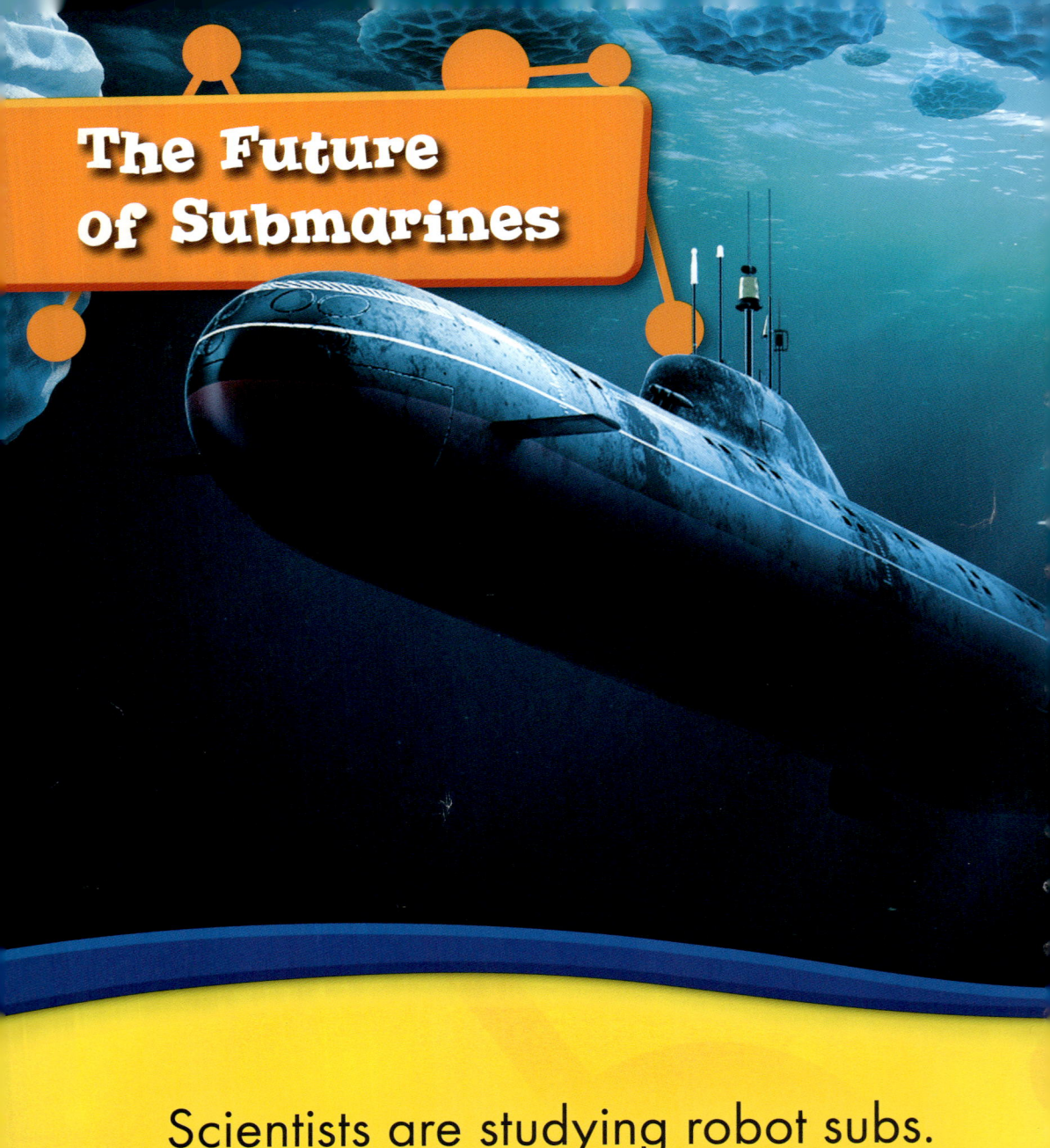

Scientists are studying robot subs. They will not need a crew. People will control them from far away. This will make subs safer.

Question

What do you think future submarines will look like?

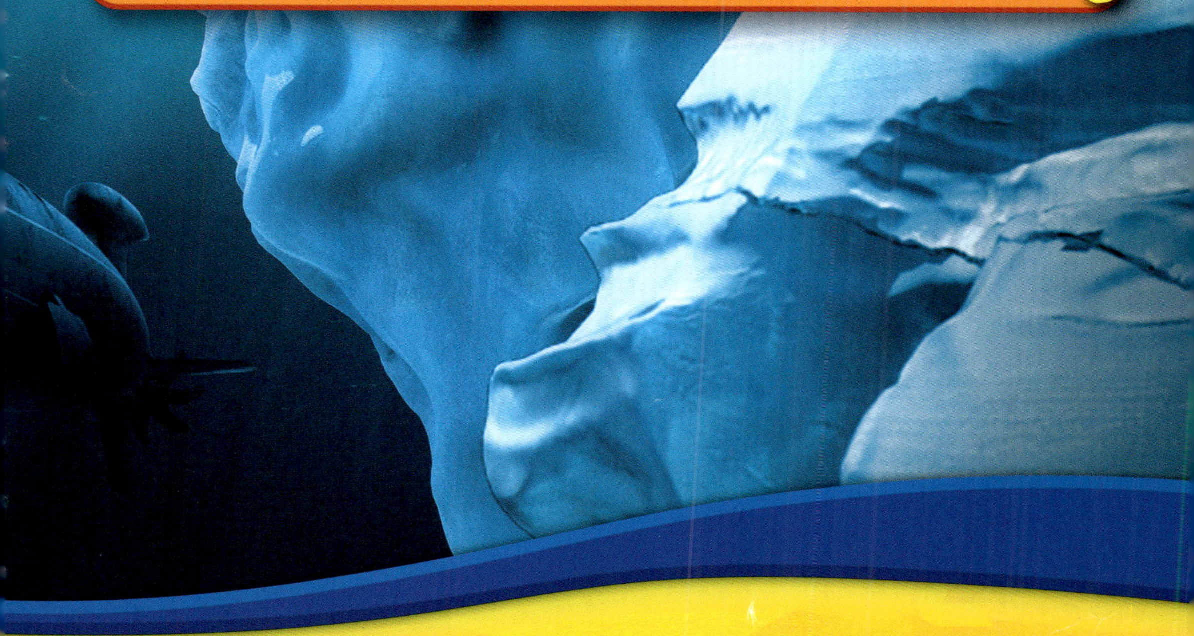

Future subs will be able to dive deeper and stay underwater longer!

Glossary

ballast tanks—large parts on submarines that hold water or air

battery—the part of a submarine that powers the sub when the engines are not running

communicates—sends and receives information

crew—the people who run a submarine

dense—having parts very close together with little space in between

diving planes—the parts on a submarine that make it move up and down

hulls—the inner and outer bodies of submarines

periscope—a long tube that sticks up from a sub's sail; the periscope is used to see above the surface of the water.

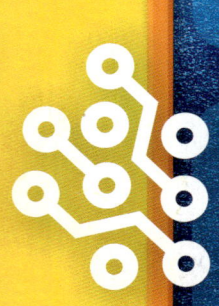

pressure—the weight or force that something makes when it pushes against something else

propeller—the part of a sub with blades that spin; the propeller helps the sub move through water.

rudders—parts of a submarine that make the sub turn left and right

sail—a tower-like part on a submarine that holds the periscope and other parts

sonar—a system that uses sound waves to find things underwater

To Learn More

AT THE LIBRARY

Brody, Walt. *How Submarines Work*. Minneapolis, Minn.: Lerner Publications, 2019.

Dellaccio, Tanya. *How a Submarine is Built*. New York, N.Y.: Gareth Stevens Publishing, 2021.

Rogers, Marie. *Incredible Submarines*. New York, N.Y.: PowerKids Press, 2022.

ON THE WEB

FACTSURFER

Factsurfer.com gives you a safe, fun way to find more information.

1. Go to www.factsurfer.com.

2. Enter "submarines" into the search box and click 🔍.

3. Select your book cover to see a list of related content.

Index

air, 14, 17
ballast tanks, 14, 16, 17
battery, 8, 9, 10, 11
communicates, 13
crew, 7, 13, 18, 19, 20
dense, 14, 16
dive, 4, 16, 21
diving planes, 10, 11
engines, 8, 10
float, 14
future, 21
hulls, 6, 7
jobs, 4
military, 4, 5
parts, 7, 9
periscope, 12, 13
pressure, 7
propeller, 10, 11
question, 21
robot subs, 20
rudders, 10, 11
sail, 12
scientists, 4, 20
sink, 16
sonar, 18, 19
sound waves, 19
steer, 11
surface, 8, 13
water, 4, 6, 7, 10, 11, 12, 14, 16, 19, 21

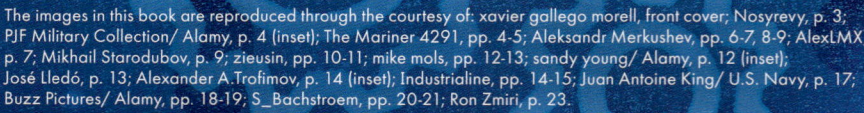

The images in this book are reproduced through the courtesy of: xavier gallego morell, front cover; Nosyrevy, p. 3; PJF Military Collection/ Alamy, p. 4 (inset); The Mariner 4291, pp. 4-5; Aleksandr Merkushev, pp. 6-7, 8-9; AlexLMX, p. 7; Mikhail Starodubov, p. 9; zieusin, pp. 10-11; mike mols, pp. 12-13; sandy young/ Alamy, p. 12 (inset); José Lledó, p. 13; Alexander A.Trofimov, p. 14 (inset); Industrialine, pp. 14-15; Juan Antoine King/ U.S. Navy, p. 17; Buzz Pictures/ Alamy, pp. 18-19; S_Bachstroem, pp. 20-21; Ron Zmiri, p. 23.